THE POETRY OF NIOBIUM

The Poetry of Niobium

Walter the Educator™

SKB

Silent King Books a WhichHead imprint

Copyright © 2023 by Walter the Educator™

All rights reserved. No part of this book may be reproduced in any manner whatsoever without written permission except in the case of brief quotations embodied in critical articles and reviews.

First Printing, 2023

Disclaimer
This book is a literary work; poems are not about specific persons, locations, situations, and/or circumstances unless mentioned in a historical context. This book is for entertainment and informational purposes only. The author and publisher offer this information without warranties expressed or implied. No matter the grounds, neither the author nor the publisher will be accountable for any losses, injuries, or other damages caused by the reader's use of this book. The use of this book acknowledges an understanding and acceptance of this disclaimer.

"Earning a degree in chemistry changed my life!"
- Walter the Educator

dedicated to all the chemistry lovers, like myself, across the world

CONTENTS

Dedication v

Why I Created This Book? 1

One - Oh, Niobium 2

Two - Forever Shine 4

Three - Wondrous Element 6

Four - Symbol Of Progress 8

Five - Side By Side 10

Six - Niobium's Allure 12

Seven - Captivating Sight 14

Eight - Science And Human Art 16

Nine - Element Rare 18

Ten - Radiant Gleam 20

Eleven - Beauty And Might 22

Twelve - Guardian Unseen 24

Thirteen - Symbol Forever	. . . . .	26
Fourteen - Through And Through	. . . . .	28
Fifteen - Power To Transform	. . . . .	30
Sixteen - Magnetic Grace	. . . . .	32
Seventeen - Exploring Galaxies	. . . . .	34
Eighteen - Satellites To Superconductors	. .	36
Nineteen - Metal So Rare	. . . . .	38
Twenty - Stardust Traveler	. . . . .	40
Twenty-One - Day And Night	. . . . .	42
Twenty-Two - Every Day	. . . . .	44
Twenty-Three - Forever Transforms	. . . . .	46
Twenty-Four - No Matter How Far	. . . . .	48
Twenty-Five - Now And Forever	. . . . .	50
Twenty-Six - Wondrous Ways	. . . . .	52
Twenty-Seven - Eternal Light	. . . . .	54
Twenty-Eight - Healing So True	. . . . .	56
Twenty-Nine - Scientific Pearl	. . . . .	58
Thirty - Changes Our World	. . . . .	60
Thirty-One - Sculptures And Jewelry	. . . .	62
Thirty-Two - Breaking Old Chains	. . . . .	64

Thirty-Three - Boundary It Breaks 66

Thirty-Four - Beyond Belief 68

Thirty-Five - Catalyst For Dreams 70

About The Author 72

WHY I CREATED THIS BOOK?

Creating a poetry book about the chemical element of niobium was an intriguing and unique endeavor. Niobium, a transition metal with atomic number 41, possesses various properties that can inspire creative expression. Poetry has the power to evoke emotions, tell stories, and explore the beauty of the world around us. By focusing on niobium, I can delve into its physical and chemical characteristics, its historical significance, and even its metaphorical representations. The challenge lies in transforming scientific information into captivating verses that engage readers and offer a fresh perspective on this lesser-known element.

ONE

OH, NIOBIUM

In the realm of elements, a noble tale unfolds,
Of a metal called niobium, its story yet untold.
With strength and resilience, it captures our gaze,
A radiant presence, that never fails to amaze.

Born in the heart of stars, forged in cosmic fire,
Niobium, the embodiment of nature's grand desire.
Its atomic number forty-one, a mark of its might,
A lustrous, silvery beauty, gleaming in the light.

Within its crystalline structure, mysteries reside,
A conductor of electricity, where currents coincide.
With superconducting properties, it defies the norm,
Unlocking doors to new realms, a scientific swarm.

In alloys and compounds, niobium finds its worth,
Enhancing strength and durability, a treasure trove unearthed.

From jet engines soaring through the sky,
To surgical implants, where healing draws nigh.

Niobium, an element of wonder and might,
Igniting innovation, in the fields of science's sight.
From laboratories to industries, it plays a vital role,
A catalyst for progress, enriching the world as a whole.

Oh, niobium, you dance with atoms unseen,
A symbol of ingenuity, in a world evergreen.
May your legacy endure, as time goes on,
A testament to the wonders of creation, forever strong.

TWO

FOREVER SHINE

In the realm of elements, a wonder we find,
A lustrous, silvery beauty, one of a kind.
Niobium, they call it, a name so refined,
With secrets and stories, forever entwined.

Oh, Niobium, you defy the norm,
With superconducting properties, you perform.
In the realm of science, you take on the form
Of a catalyst for progress, a beacon so warm.

From jet engines soaring, through the sky,
To surgical implants, where strength does lie,
Niobium, you enhance, you fortify,
In industries vast, where dreams they imply.

A symbol of wonder, you ignite the flame,
Unveiling the mysteries, none can tame.

In labs and experiments, you play the game,
Unleashing your power, your endless acclaim.

   Oh, Niobium, your legacy grand,
A testament to ingenuity, oh, so grand.
Through the ages, you'll forever stand,
A marvel of creation, by the creator's hand.

   So let us celebrate, this element divine,
Niobium, the jewel, where wonders intertwine.
In the realm of science, you'll forever shine,
A testament to the marvels we find.

THREE

WONDROUS ELEMENT

In the realm of elements, there lies a wonder,
A metal strong, a force to ponder.
Niobium, the name it bears,
With resilience unmatched, it dares.
    A conductor of power, a superconducting delight,
In the field of physics, it shines so bright.
At frigid temperatures, it defies,
Flowing currents without compromise.
    In the depths of the Earth, it does reside,
Extracted with care, its beauty can't hide.
From the mines it emerges, raw and pure,
Ready to shape the world, to endure.
    In the realm of industry, it finds its place,
Aiding progress at an incredible pace.

From aerospace to medicine, it plays a part,
Niobium's versatility, a work of art.
  It strengthens steel, it lightens the load,
In bridges and buildings, a sturdy abode.
Its alloys bring strength, its presence profound,
A symbol of progress, a marvel renowned.
  Niobium, the element of dreams,
Igniting a fire, breaking old regimes.
In the world of science, it takes flight,
Pushing boundaries, shining bright.
  A testament to innovation and might,
Niobium, a star in the chemical night.
Oh, wondrous element, so rare and true,
In our quest for knowledge, we turn to you.

FOUR

SYMBOL OF PROGRESS

In the realm of elements, a star so bright,
Niobium, a metal of wondrous might.
With strength and resilience, it does ignite,
A symphony of wonders, shining so light.

In superconductivity, it finds its grace,
A conductor of energy, at an incredible pace.
Magnetic fields it defies, with an elegant embrace,
Unlocking secrets of science, in this cosmic chase.

From the depths of the Earth, it does arise,
A treasure hidden, beneath azure skies.
In industry it thrives, a versatile prize,
Forging a future, where progress never dies.

In steel it imparts, a strength so true,
A guardian against forces, that threaten to undo.
An ally in construction, staunch and anew,
With Niobium, dreams and visions come into view.

Oh Niobium, you shimmer and shine,
A beacon of wonder, so divine.
Symbol of progress, innovation, and time,
In the realm of elements, you truly shine.

FIVE

SIDE BY SIDE

In the realm of elements, shining bright,
A metal named Niobium, a wondrous sight.
With atomic number forty-one, it stands tall,
A symbol of science, captivating all.

Niobium, a lustrous and silvery gleam,
A metal so strong, a chemist's dream.
In alloys it's found, enhancing their might,
Improving strength, making them take flight.

From bridges to pipelines, so many things,
Niobium's presence, endurance it brings.
Its resistance to corrosion, a remarkable trait,
In industry and engineering, it's first-rate.

In superconductivity, it plays a key role,
At temperatures low, it takes a mighty toll.

Unlocking new frontiers, pushing boundaries far,
Niobium's power, like a shining star.

And in the realm of physics, it's a force to behold,
With its magnetic properties, a story untold.
In magnets and accelerators, its presence is grand,
Unleashing energy, pushing science's hand.

Oh, Niobium, a metal of strength and grace,
Innovation and progress, it does embrace.
A symbol of resilience, it stands with pride,
Forever pushing boundaries, side by side.

SIX

NIOBIUM'S ALLURE

In the realm of elements, Niobium stands tall,
A metal of strength, it captivates us all.
With atomic number forty-one, it shines bright,
A star in the periodic table's grand sight.

In aerospace, it soars through the sky,
Niobium's alloys, they never comply.
With wings that carry dreams to new heights,
It enhances strength, ensuring enduring flights.

Medicine finds solace in Niobium's might,
Implants and prosthetics, it brings to light.
With biocompatibility, it heals and mends,
Aiding the broken, as hope transcends.

Now, let's explore another facet of this treasure,
Niobium, a companion in structures of pleasure.

In bridges and buildings, where weight is a strain,
It lightens the load, ensuring they sustain.

In steel, it bestows resilience and grace,
Defying the elements, standing firm in its place.
With strength and durability, it never fails,
A testament to Niobium's unwavering tales.

Now, let us delve into a wondrous sight,
Where Niobium dances in the realm of light.
Superconductivity, a marvel to behold,
Niobium defies magnetic fields, bold.

In wires and magnets, it conducts with ease,
Powering the future, with innovation's keys.
A realm where science and magic align,
Niobium's prowess, forever divine.

In industry, engineering, and physics profound,
Niobium's resistance to corrosion, renowned.
A guardian against time's relentless decay,
It stands as a symbol, come what may.

So, let us celebrate Niobium's allure,
A versatile element, steadfast and pure.
In every industry, it leaves its mark,
A testament to its brilliance, shining in the dark.

SEVEN

CAPTIVATING SIGHT

In realms unseen, where wonders lie,
There dwells a force that catches the eye.
Niobium, the element of might,
A shining star in the darkest night.

In steel's embrace, it lends its hand,
Forging bonds that forever stand.
From bridges tall to ships that sail,
Niobium's strength will never fail.

In jet engines roaring through the sky,
Niobium soars, it learns to fly.
Its wings of metal, light and strong,
Defying gravity all day long.

In medicine's realm, a healing touch,
Niobium aids, it means so much.
Implants and devices, life restored,
It truly is a precious reward.

In laboratories, where science thrives,
Niobium sparks, as knowledge arrives.
Its conductive nature, a gift untold,
Unraveling mysteries yet to unfold.

A guardian against time's cruel grip,
Niobium resists corrosion's whip.
A shield of beauty, eternally bright,
Defying decay with all its might.

And in the realm of magnetism's call,
Niobium stands tall, above them all.
Its magnetic dance, a captivating sight,
Guiding our world into the light.

So let us raise a toast to thee,
Niobium, the element that sets us free.
A symbol of progress, resilience, and more,
Forever engraved in the world's grand lore.

EIGHT

SCIENCE AND HUMAN ART

In realms of strength, a metal bold,
Lies Niobium, a tale untold.
With steadfast grace, it takes its stand,
A master of the elements, hand in hand.

In bridges towering, it lends its might,
A sturdy shield against the night.
In skyscrapers reaching for the sky,
It holds the dreams that soar up high.

In the heart of engines, it beats with fire,
A force untamed, a burning desire.
From planes that soar to cars that race,
It fuels the dreams we dare to chase.

In wires coiled, a dance of might,
Conducting power, a brilliant light.

In circuits humming, it finds its place,
Guiding the current with elegant grace.
    In laboratories, its secrets unfold,
Unlocking the mysteries, untold.
In physics and superconductivity,
It reveals the wonders of electricity.
    In magnets strong, it pulls the tide,
A force magnetic, it cannot hide.
In fusion's quest, it holds the key,
To powers unseen, a future to be.
    Oh Niobium, a metal rare,
A symbol of strength beyond compare.
In every realm, you play your part,
A testament to science and human art.

NINE

ELEMENT RARE

In the realm of elements, a star does shine,
A metal that's noble, a treasure divine.
Its name is Niobium, a marvel untold,
A world of wonders, its secrets unfold.

In laboratories, where science takes flight,
Niobium dances, bathed in pure light.
Its atomic number, a magical sign,
It's the pride of the periodic table, this metal fine.

In the realm of aviation, it soars to the sky,
Crafting wings of strength, that are destined to fly.
In engines and turbines, it prevails with might,
Enduring the heat, pushing boundaries with delight.

In the realm of medicine, it brings healing grace,
Implants and devices, enhancing life's pace.

With its biocompatible soul, it mends broken hearts,
A guardian warrior, playing its part.
    In the realm of art, it's a sculptor's delight,
A canvas for creatives, a symphony of light.
Its luster and shine, a testament true,
Niobium's beauty, forever anew.
    So let us celebrate this element rare,
With its versatility and strength, beyond compare.
Niobium, the metal of dreams, we adore,
Forever it shall shine, forever it shall endure.

TEN

RADIANT GLEAM

In the realm of science, a marvel there lies,
A metal so wondrous, that it defies,
Corrosion and rust, it resists with great might,
Niobium, a beacon, shining so bright.

Its atoms entwined, in a crystalline dance,
Forming a lattice, with elegance and grace,
A conductor of currents, so pure and so true,
Superconductivity, it brings to the view.

From magnets it harnesses, a mystical power,
Creating fields strong, like a mythical tower,
In MRI machines, it aids in our quest,
To heal and to diagnose, with accuracy blessed.

Niobium, oh niobium, a marvel to behold,
In jet engines it soars, with a strength untold,

Its alloys, so versatile, in bridges and beams,
Supporting our structures, fulfilling our dreams.

But beyond its utility, a beauty resides,
Adorning our jewelry, with shimmer and glides,
A lustrous reflection, a radiant gleam,
Niobium, the element, in a poet's dream.

So raise your glass, to this element grand,
Niobium, the hero, of industry's band,
A symbol of progress, a promise of more,
With each passing day, unlocking new doors.

ELEVEN

BEAUTY AND MIGHT

In the realm of metals, a gem so rare,
Niobium shines with an elegant flair.
With strength and grace, it stands tall and true,
A symbol of resilience, in all that it can do.

Resistance to corrosion, a marvelous feat,
Niobium withstands the elements it must meet.
In the face of time, it remains steadfast,
A guardian of progress, meant to forever last.

Oh, Niobium, conductor of dreams,
Your superconductivity, a marvel it seems.
Unlocking the secrets of energy's flow,
You light up the world with a radiant glow.

Magnetic whispers dance in your core,
Drawing us closer, forevermore.

In the depths of your essence, mysteries unfold,
A magnetic force, a story yet untold.
From industries vast to the artist's hand,
Niobium's touch, a stroke so grand.
In medicine's embrace, you heal and mend,
A silent hero, a helping hand to lend.
Niobium, a metal of beauty and might,
A symphony of elements, harmoniously bright.
In your presence, we find strength and grace,
A testament to progress, in every embrace.

TWELVE

GUARDIAN UNSEEN

In the realm of elements, a wonder does arise,
Niobium, the metal, with its captivating guise.
Through the ages, it has stood, unwavering and strong,
A symbol of resilience, where progress does belong.

With conductive prowess, it dances with the flow,
Harnessing energy, where currents freely go.
A conductor of dreams, it lights up the way,
In circuits and wires, where technology holds sway.

Corrosion, it defies, with a steadfast will,
Unyielding to rust, it remains polished and still.
Through rain and storm, it stands tall and gleams,
A shield of protection, as nature's wrath teems.

Magnetic in nature, it draws forces unseen,
In magnets and alloys, a magnetic dream.

Attracting and repelling, a dance in the air,
Niobium's allure, a magnetic affair.

In medicine's realm, it finds a noble place,
In implants and devices, it supports life's race.
A healer's companion, a guardian unseen,
Niobium, the lifeline, where miracles convene.

In art's grand canvas, it lends its grace,
A sculptor's delight, a masterpiece to embrace.
In jewelry and adornments, it adds a touch divine,
Niobium's beauty, forever to shine.

Oh, Niobium, the element of might,
A symbol of progress, a beacon of light.
In science and industry, you hold the key,
To a future bright, where possibilities are free.

THIRTEEN

SYMBOL FOREVER

In the depths of the Earth, a secret lies,
A metal born of fire, with hidden ties.
Niobium, the element of strength and might,
With conductive powers that shine so bright.

Resistance to corrosion, it stands unyielding,
A shield against the elements, forever revealing.
In industries, it molds and shapes,
Forging progress, with no escapes.

Magnetic properties, a force to reckon,
Drawing us closer, like a magnetic beckon.
In medicine's embrace, it brings healing,
Unlocking mysteries, the heart's revealing.

Versatile in art, a canvas to explore,
Crafting beauty, like never before.

Adorned in jewelry, it gleams with grace,
A symbol of elegance, in every embrace.
    Niobium, a metal, a marvel so grand,
A testament to resilience, built to withstand.
Unleashing potential, in every endeavor,
A catalyst for dreams, a symbol forever.
    So let us celebrate, this element divine,
A bridge to the future, where possibilities shine.
Niobium, a beacon of progress and might,
Guiding us forward, with a radiant light.

FOURTEEN

THROUGH AND THROUGH

In the realm of wonders, a metal resides,
Niobium, with strength that defies,
A silent hero, hidden from sight,
Unveiling mysteries, with pure light.

In art and industry, its role profound,
A conductor of dreams, with secrets unwound,
From skyscrapers soaring, reaching the sky,
To bridges that span, where rivers run high.

In medicine's embrace, it lends a hand,
Implants of hope, where healing is planned,
A guardian of health, with remarkable grace,
Mending broken bodies, leaving no trace.

In technology's realm, it sparks innovation,
Pushing boundaries, with boundless creation,

From superconductors, with power untamed,
To microchips humming, where knowledge is gained.
   With beauty unmatched, it dazzles the eye,
A palette of colors, no painter can deny,
From jewelry to sculptures, it's a sight to behold,
A testament to grace, more precious than gold.
   Oh, Niobium, you unlock the door,
To a future unbounded, where dreams can soar,
With strength and resilience, you pave the way,
To a world filled with promise, a brighter day.
   So let us celebrate, this element true,
Niobium, the shining star, through and through,
A symbol of progress, a beacon of light,
Guiding us forward, with all of our might.

FIFTEEN

POWER TO TRANSFORM

In the realm of steel and fire, a gleaming jewel does reside,
A beacon of strength and promise, with secrets it does hide.
Niobium, oh noble element, your story yet untold,
A symphony of atoms, a tale of the bold.

In the halls of medicine, your touch brings healing grace,
Implants of resilience, where life finds its embrace.
Bones made stronger, hearts repaired, a testament to your might,
Niobium, the guardian, guiding us through the night.

In art's enchanting realm, your colors dance and sing,
A palette of endless possibilities, creativity takes wing.

From jewelry to sculptures, your beauty does inspire,
Niobium, the muse, kindling passion's fire.

In the realm of technology, you shape our modern age,
Superconductors and alloys, turning the page.
From satellites in space to the engines on the ground,
Niobium, the innovator, with progress tightly bound.

Oh Niobium, a symbol of resilience and light,
A bridge between the past and future, shining ever bright.
In your atoms, we find hope, in your presence, we find strength,
Niobium, the catalyst, forging a world of extended length.

So let us raise our voices, and sing your praises high,
Niobium, oh noble element, in you, we shall rely.
For in your essence, lies the power to transform,
A testament to human spirit, in harmony we'll conform.

SIXTEEN

MAGNETIC GRACE

In the realm of elements, Niobium shines,
A versatile metal with secrets untold.
Its presence in industries, a testament sublime,
To the wonders of science, forever unfold.

In the depths of alloys, it lends its might,
Strengthening structures with its resilient core.
From bridges to airplanes, it holds them tight,
A guardian of progress, forevermore.

In magnets, it dances with magnetic grace,
Attracting and repelling with a mysterious force.
Unlocking the secrets of the universe's embrace,
Revealing the wonders of nature's discourse.

In the realm of medicine, it plays its part,
Implants of Niobium, a gift of healing touch.
Mending broken bones, mending a fragile heart,
An ally in health, we cherish it so much.

In art's vast canvas, it paints a gleam,
Adorning sculptures with a radiant hue.
Transforming the ordinary into a vivid dream,
A symbol of beauty, forever anew.

Oh, Niobium, element of wonder and awe,
Your contributions, a testament to human quest.
A beacon of hope, a glimpse of what's in store,
In your hands, a brighter future will manifest.

SEVENTEEN

EXPLORING GALAXIES

In the realm of secrets untold,
Lies a metal, resilient and bold.
Niobium, the enigmatic star,
Unveiling mysteries from afar.

In medicine's embrace, it gleams,
A healer in the realm of dreams.
With strength and grace, it mends the weak,
A soothing balm for those who seek.

From canvas to sculpture, art comes alive,
Niobium's touch, a creative drive.
Its hues dance with vibrant delight,
A kaleidoscope of colors, shining bright.

In technology's realm, it takes flight,
A catalyst for progress, shining so bright.

Unlocking doors to a boundless land,
Niobium's touch, a visionary's hand.

    A cosmic voyager, it joins the quest,
Exploring galaxies, the very best.
Unraveling the universe's tapestry,
Niobium's grace, a celestial symphony.

    Oh, Niobium, you inspire and ignite,
A beacon of hope, shining so bright.
Resilient and wondrous, a gift to behold,
Your potential, a story yet untold.

EIGHTEEN

SATELLITES TO SUPERCONDUCTORS

In the realm of metals, a marvel is found,
A treasure of strength, with beauty unbound.
Niobium, the element, steadfast and true,
A symphony of atoms, in shades of blue.

Behold its wonders, so versatile and grand,
A beacon of progress, across the land.
In art, it dances, with colors it weaves,
A canvas of dreams, where imagination believes.

In medicine's grip, it extends its hand,
A healer in disguise, a remedy so grand.
With stents and implants, it mends broken hearts,
Restoring the rhythm, where life's beat restarts.

In the realm of technology, it takes its flight,
A conductor of currents, igniting the light.

From satellites to superconductors it soars,
Unleashing the power, that innovation adores.

In the cosmos above, where stars brightly gleam,
Niobium whispers, a celestial theme.
For in supernovas, where elements collide,
It forges the stars, where galaxies reside.

Oh Niobium, a catalyst divine,
Unleashing potential, where boundaries untwine.
With grace and resilience, it marches ahead,
A symbol of progress, where dreams are spread.

So let us celebrate this noble element,
With poems and songs, in its honor, we're sent.
For Niobium, the wonder, so rare and true,
A testament to possibilities anew.

NINETEEN

METAL SO RARE

In the depths of Earth, where secrets lie,
A metal emerges, gleaming, shy.
Niobium, the element of might,
Unveiling wonders, casting out the night.

In laboratories, it takes its stand,
A catalyst of progress, a helping hand.
From medicine's grasp to art's embrace,
Niobium weaves dreams with gentle grace.

Its atomic dance, a symphony grand,
Unleashing potential, a master plan.
A conductor of change, resilient and true,
Niobium shines, a beacon through.

In the cosmos above, where stars collide,
Niobium journeys, on a celestial ride.
A bridge between worlds, an interstellar guide,
Unlocking mysteries, where wonders reside.

From microchips to turbines' roar,
Niobium fuels innovation's core.
A foundation of strength, unyielding and vast,
Building a future, destined to last.

Oh Niobium, element divine,
In you, hope and progress intertwine.
With your shimmering gleam, a promise we see,
That through your power, our world can be free.

So let us embrace, this metal so rare,
And revel in possibilities, beyond compare.
Niobium, the key to a brighter day,
A symbol of hope, lighting our way.

TWENTY

STARDUST TRAVELER

In the realm of atoms, a wondrous might,
There lies a metal, with a gleaming light.
Niobium, the element, so pure and strong,
Unleashing potential, where dreams belong.

In laboratories, its secrets unfurled,
A catalyst for progress, in a changing world.
With every discovery, a new door opens wide,
Niobium's touch, a spark, a guide.

In the realm of technology, it takes its stance,
Woven in circuits, enhancing every chance.
From smartphones to spacecraft, it weaves its spell,
Niobium's magic, where innovation dwells.

But beyond the realm of wires and chips,
Niobium dances through the artist's fingertips.

In hues of iridescence, its canvas takes flight,
Inspiring beauty, in the darkest night.

A bridge between science and the healing art,
Niobium lends its grace, playing a vital part.
In medical marvels, its presence shines bright,
Bringing hope and healing, an ethereal light.

And in the depths of the cosmos, it soars,
A stardust traveler, exploring distant shores.
Niobium's essence, in celestial skies,
Ignites our imagination, as it mesmerizes.

Oh, Niobium, element of wonder and grace,
Unveiling mysteries, in every embrace.
A symbol of progress, a beacon of hope,
In your atomic embrace, we find strength to cope.

So let us celebrate, this metal divine,
Niobium, the catalyst, the radiant sign.
For in its presence, our spirits take flight,
Unleashing potential, in a world burning bright.

TWENTY-ONE

DAY AND NIGHT

In the realm of dreams, a metal gleams,
A marvel woven in nature's seams,
Niobium, a cosmic star,
Unveiling secrets from afar.

With strength and grace, it takes the stage,
A catalyst for the modern age,
In superconductors, it ignites,
Unleashing power, defying nights.

From medicine's embrace, it lends a hand,
Implants of hope, a promise grand,
Mending bones, healing hearts,
Niobium's touch, a work of art.

In hues of blue and silver's glow,
Its beauty shines, a vibrant show,

A symbol of resilience, strong and true,
Niobium, we turn to you.

Exploring space, it reaches far,
Unleashing wonders, like a shooting star,
From satellites to telescopes high,
Niobium guides us, beyond the sky.

In every atom, a story unfolds,
Of endless possibilities, yet untold,
A metal, humble, yet so profound,
Niobium, where wonders abound.

So let us celebrate its might,
A beacon in the darkest night,
Niobium, a shining light,
Unveiling wonders, day and night.

TWENTY-TWO

EVERY DAY

In the realm of science and exploration,
There lies a metal of profound creation.
Niobium, the element of wonder,
In its essence, secrets it does encumber.

In the realm of medicine, it holds its sway,
Aiding healers on their noble way.
Implants and prosthetics, strong and true,
Niobium's strength, bringing life anew.

In the cosmos, it ventures, bold and bright,
Aiding telescopes in their celestial flight.
Observing stars in their cosmic dance,
Niobium lenses, capturing each glance.

In the realm of technology, it takes the lead,
Powerful and versatile, fulfilling every need.

Superconductors, defying resistance,
Niobium's magic, unlocking persistence.
  But beyond the realm of science and steel,
Niobium's beauty reveals its appeal.
In art and jewelry, it shines so bright,
Bringing vibrant colors, a mesmerizing sight.
  Oh, Niobium, element of grace,
In every realm, you find your place.
From medical miracles to cosmic sights,
Your presence brings wonder and delights.
  So, let us celebrate this element rare,
With awe and wonder, let's be aware.
For Niobium, in its unique way,
Gifts us with hope, each and every day.

TWENTY-THREE

FOREVER TRANSFORMS

In the realm of currents, a conductor stands tall,
Niobium, the name that echoes through it all.
A symbol of power, a catalyst of might,
Unleashing innovation, igniting the light.

In jewelry's embrace, it finds its place,
Anodization's magic, colors with grace.
Resilient and pure, it shines with allure,
A testament to beauty, forever to endure.

From Large Hadron Collider to MRI's gaze,
Niobium's alloys in scientific arrays.
Superconducting magnets, a marvel to behold,
Unveiling secrets, mysteries untold.

In gas pipelines and engines, it lends its strength,
In aerospace marvels, it reaches great length.

A steadfast companion, through fire and air,
Fueling progress, building a future fair.
But beyond the earthly realm, its wonders transcend,
A connection to cosmos, where stars ascend.
Forging celestial bodies, an alchemist of space,
Niobium's presence, a celestial embrace.
So let us marvel at this element divine,
A symbol of hope, a promise to shine.
In realms of science, art, and dreams,
Niobium's presence, a symphony it seems.
From healing in medicine, a beacon of light,
To technology's realm, a vision so bright.
Versatile and wondrous, in all its forms,
Niobium, the element that forever transforms.

TWENTY-FOUR

NO MATTER HOW FAR

In the realm of medicine, Niobium shines,
A silent hero in medical design.
Implants and devices, it does create,
To heal and restore, to alleviate.

Within our bodies, it finds its way,
A catalyst for progress, day by day.
Implanted deep, where healing resides,
Niobium's touch, a gift that provides.

Beyond our world, in the cosmos it dwells,
A cosmic explorer, where mystery swells.
In telescopes and satellites, it's found,
Unveiling secrets, with every profound.

But let us not forget its artistic grace,
In jewelry and art, it finds its place.

A metal of beauty, shining in hues,
Adorning our lives, with colors so true.
 Versatile and strong, Niobium stands,
In technology's realm, it commands.
A conductor, a superconductor of power,
Pushing the limits, each passing hour.
 Niobium, a symbol of hope and light,
Guiding us forward, through darkness and fright.
In medicine, art, and technology's hold,
A beacon of progress, forever bold.
 So let us celebrate this element's might,
For its contributions, day and night.
Niobium, a hero, a guiding star,
Leading us onward, no matter how far.

TWENTY-FIVE

NOW AND FOREVER

In labs of science, Niobium resides,
A metal of wonder, where progress abides.
In medicine's realm, it plays a key role,
Implants and devices, mending the soul.

Technology's realm, it's there to be seen,
In capacitors, wires, and screens that gleam.
With superconductors, it dances in grace,
Unleashing potential, at an incredible pace.

In art's vast canvas, its beauty it lends,
A palette of colors, where creativity blends.
From sculptures to jewelry, it shines bright,
Captivating hearts, in its radiant light.

Niobium, versatile, oh what a delight,
A beacon of hope, in the darkest of night.

With wings of progress, it soars through the sky,
In aerospace's realm, where dreams multiply.
   Beyond our Earth's bounds, to the cosmos it flies,
Exploring the unknown, where stars mesmerize.
A bridge to the heavens, where wonders unfold,
Niobium, the element, a story untold.
   So let us rejoice, in Niobium's might,
For in its presence, there's endless delight.
From science to art, it brings us together,
A symbol of progress, now and forever.

TWENTY-SIX

WONDROUS WAYS

In the realm of science's embrace,
A metal of marvel and grace,
Niobium, a radiant light,
Guides humanity's eternal flight.

In medicine's healing touch,
Niobium's power, it does clutch,
Mending bones, soothing pain,
A guardian against life's strain.

In technology's vibrant domain,
Niobium's prowess does reign,
Circuits of wonder, devices of might,
Unleashing progress, day and night.

In art's creative tapestry,
Niobium weaves its legacy,

Brushstrokes of color, forms divine,
Crafting beauty, a sight to define.

In aerospace's lofty quest,
Niobium soars with zest,
Beyond the stars, to the unknown,
Unveiling mysteries, uncharted zones.

A catalyst of dreams, it be,
Niobium, the key to see,
The boundless potential that lies within,
A source of hope, where wonders begin.

From laboratory to workshop's gleam,
Niobium's presence, a vibrant seam,
Inspiring minds, igniting the fire,
Unleashing passions, lifting us higher.

Oh, Niobium, element of might,
A beacon in the darkest night,
We hail your power, we sing your praise,
Forever grateful for your wondrous ways.

TWENTY-SEVEN

ETERNAL LIGHT

In the realm of science's embrace,
Where dreams and wonders interlace,
Lies a gem that shines with grace,
Niobium, a star in space.

With strength and beauty, it does ignite,
A beacon in the darkest night,
A bridge that spans the great divide,
In art and tech, it does preside.

In circuits, it conducts the flow,
A symphony of electrons in tow,
Empowering devices, high and low,
Niobium's magic, a vibrant glow.

In wings of steel, it takes to flight,
Aerospace dreams reaching new heights,
Exploring the heavens, vast and wide,
With Niobium, we dare to ride.

In medicine's hands, it brings relief,
Implants of hope, a precious belief,
Mending broken hearts, a soothing touch,
Niobium's healing, hearts aflame, as such.

Oh, Niobium, a catalyst of dreams,
Unleashing potential, in vibrant streams,
A symbol of progress, hope, and might,
Guiding us forward, in its eternal light.

TWENTY-EIGHT

HEALING SO TRUE

In the realm of aerospace, where dreams take flight,
There soars a metal, shining ever so bright.
Niobium, the star of the celestial show,
With strength and grace, it helps us reach new heights.

In engines roaring, it dances with the flame,
Withstanding heat, it refuses to tame.
A guardian of the skies, a protector of flight,
Niobium, our ally, in the darkest of nights.

But beyond the clouds, where artistry resides,
Niobium unveils its elegant guise.
In jewelry it gleams, a precious treasure,
Adorning the necks of those who seek pleasure.

Its lustrous hues, a painter's delight,
Niobium, a muse, in shades so bright.
A canvas of expression, a sculptor's delight,
Niobium, the essence of creativity's might.

In medicine's realm, where healing is sought,
Niobium lends a hand, a cure it has brought.
Implants that mend, with strength and precision,
Niobium, the healer, with a gentle vision.

In technology's grasp, where progress takes hold,
Niobium's versatility, a story untold.
From electronics to superconductors sublime,
Niobium, the innovator, transcending time.

And in the realm unknown, where mysteries lie,
Niobium's power, it helps us defy.
Exploring the depths, unraveling the unknown,
Niobium, the explorer, a guide to the throne.

Oh, Niobium, a symbol of hope and might,
In every field, you bring forth the light.
Empowering devices, with healing so true,
Niobium, we celebrate the wonders of you.

TWENTY-NINE

SCIENTIFIC PEARL

In the realm of space, where stars collide,
A metal so mighty, it takes us for a ride.
Niobium, the element of awe and grace,
Unveils its wonders in cosmic embrace.

Through the skies, where rockets soar,
Niobium crafts the vessels that explore.
From Earth to distant realms, it takes flight,
Guiding us to the edges of the night.

In laboratories, its secrets unfurl,
A versatile element, a scientific pearl.
From alloys strong to superconducting might,
Niobium brings progress and healing to light.

In artistry, it finds its place,
Adorning jewels with its gleaming grace.
In colors vibrant, it dances and gleams,
A symbol of beauty, beyond our wildest dreams.

In technology's realm, it takes the lead,
Pushing boundaries, fulfilling our need.
From superconductors to high-tech devices,
Niobium's power, the world entices.

Oh, Niobium, element divine,
Connecting our world with the cosmic line.
Inspiring creativity, forging paths anew,
A symbol of hope, progress, and breakthrough.

So let us celebrate this mighty metal's worth,
For Niobium, the element that brings us forth.
In the cosmos, in art, in science's quest,
Niobium shines, the element that's best.

THIRTY

CHANGES OUR WORLD

In the realm of elements, a treasure untold,
There lies a metal, vibrant and bold.
Niobium, the beauty that nature has made,
With wonders and secrets, it's willing to trade.

In science and industry, it finds its home,
A versatile force, through time it has roamed.
From technology's grip to aerospace's flight,
Niobium shines, a beacon of light.

In medicine's realm, it brings healing and grace,
An ally in battles that we all face.
Exploring new frontiers, it stands tall and true,
Niobium, the element that pushes us through.

But beyond the confines of practical gain,
Niobium's allure holds an artistic strain.

In jewelry and art, it gleams and it glows,
Captivating hearts with its radiant shows.

Electronics it empowers, circuits it weaves,
A conductor of energy, its power it achieves.
Innovation's catalyst, it ignites the flame,
Niobium, the element that fuels our aim.

And as we seek progress, in every endeavor,
Niobium's essence, a constant endeavor.
A symbol of hope, and dreams yet unfurled,
Niobium, the metal that changes our world.

THIRTY-ONE

SCULPTURES AND JEWELRY

In the realm of elements, a star does shine,
Niobium, its name, a tale divine.
A metal bold, with secrets to unfold,
Its wonders, a story yet untold.

In aerospace it soars, a guiding light,
With strength and lightness, taking flight.
From rockets to satellites, it takes its place,
Navigating the vastness of outer space.

In labs of science, it dances with grace,
A conductor of currents, a chemical embrace.
Superconductivity, its gift to behold,
Unleashing potential yet untold.

In art, it paints a masterpiece so rare,
A canvas of colors, beyond compare.

Sculptures and jewelry, crafted with care,
Niobium's allure, forever to wear.

    Its healing touch, a balm for the soul,
Mending the broken, making us whole.
Implants and prosthetics, a life reborn,
Niobium's embrace, a future adorned.

    In technology's realm, it paves the way,
Connecting the world, day after day.
From smartphones to laptops, its power unleashed,
Niobium, a force that will never cease.

    Oh Niobium, a metal divine,
With versatility that knows no line.
In science, in art, in technology's grace,
You shape our world, leaving a lasting trace.

THIRTY-TWO

BREAKING OLD CHAINS

In skies above, where dreams take flight,
A metal glows, with shimmering light.
Niobium, a star in the cosmic sea,
A beacon of hope, guiding you and me.

In aerospace, it spreads its wings,
Crafting vessels that soar and sing.
From rockets to planes, it takes us high,
Through boundless realms, where dreams can fly.

In medicine's embrace, it heals and mends,
A guardian of health, where hope ascends.
Implanted with care, it strengthens the weak,
A shield against pain, a remedy we seek.

In technology's realm, it sparks innovation,
Igniting progress, a source of fascination.

From circuits to magnets, it shapes our world,
Unleashing wonders, as secrets unfurled.

In art's embrace, it dazzles and gleams,
Transforming visions, beyond all extremes.
A canvas of beauty, where creativity thrives,
Niobium's touch, where imagination survives.

Throughout the ages, its presence remains,
A symbol of progress, breaking old chains.
Niobium, a hero, with powers untold,
Innovating the world, with stories yet unfold.

So let us celebrate, this element divine,
Niobium's glory, forever will shine.
With hope as its guide, it paves the way,
A metal of wonders, forever will stay.

THIRTY-THREE

BOUNDARY IT BREAKS

In the realm of elements, Niobium stands tall,
A magical metal, a wonder for all.
With healing properties, it mends the unwell,
Bringing solace and comfort, its stories they tell.

In technology's realm, Niobium shines bright,
Exploring the unknown, pushing boundaries with might.
From spacecraft to turbines, its strength it imparts,
A metal of progress, igniting new starts.

In the realm of art, Niobium is a gem,
Crafted into jewelry, a radiant diadem.
Its colors so vibrant, a kaleidoscope of hue,
A masterpiece created, captivating and true.

A symbol of hope, a beacon of light,

Niobium's touch heals and makes everything right.
In labs and in factories, it powers the way,
Advancing our world, propelling us day by day.

    Versatile and mighty, in every form it takes,
Niobium's power, no boundary it breaks.
From medicine to space, art to innovation,
It shapes and transforms, without hesitation.

    So let us celebrate Niobium's grace,
Its presence profound, its impact we embrace.
For in this metal, lies the key,
To a future of progress, a world set free.

THIRTY-FOUR

BEYOND BELIEF

In the realm of endless possibilities,
Where healing meets versatility,
Lies a metal, strong and rare,
A symbol of hope beyond compare.
    Niobium, oh radiant star,
In technology, you've traveled far.
From superconductors to capacitors,
Your wonders shine like guiding stars.
    In medicine, you bring relief,
A precious gift, beyond belief.
Implants and prosthetics, you enhance,
Restoring lives with a graceful dance.
    A conductor of electricity,
A catalyst for creativity,
In every realm, you leave your mark,
Igniting progress with a vibrant spark.

From skyscrapers reaching high,
To bridges spanning the endless sky,
Your strength and resilience, they endure,
A testament to your beauty, pure.

Niobium, a jewel of art,
A masterpiece that sets us apart.
In every brushstroke, every stroke of pen,
You inspire us to create again.

Oh, Niobium, metal divine,
In your presence, we shall shine.
In science, technology, and even art,
You touch our souls, you touch our hearts.

THIRTY-FIVE

CATALYST FOR DREAMS

In the realm of science, a jewel we find,
A metal so noble, a treasure defined.
Niobium, the element, with powers untold,
A catalyst for progress, a story to unfold.

    In the realm of healing, it takes center stage,
Implants and prosthetics, it helps to engage.
With strength and resilience, it mends broken bones,
Restoring our bodies, with grace it atones.

    In the realm of conductivity, it shines so bright,
A conductor of electrons, a beacon of light.
In wires and cables, it sparks innovation,
Igniting the fires of progress, with no hesitation.

    In the realm of art, it's a canvas to explore,
A medium of beauty, we truly adore.

Its hues and textures, a feast for the eyes,
Inspiring creations, that touch the skies.

In the realm of imagination, it knows no bounds,
A catalyst for dreams, where creativity resounds.
In science, technology, and art it plays its part,
Touching our souls, igniting our hearts.

Niobium, oh marvel, so versatile and strong,
A symbol of endurance, where dreams belong.
In every field it leaves its mark,
A spark of life, a glowing arc.

So let us celebrate this noble element,
A jewel of art, with no precedent.
In science, technology, and art it weaves,
A tapestry of wonder, that forever relieves.

ABOUT THE AUTHOR

Walter the Educator is one of the pseudonyms for Walter Anderson. Formally educated in Chemistry, Business, and Education, he is an educator, an author, a diverse entrepreneur, and he is the son of a disabled war veteran. "Walter the Educator" shares his time between educating and creating. He holds interests and owns several creative projects that entertain, enlighten, enhance, and educate, hoping to inspire and motivate you.

> Follow, find new works, and stay up to date
> with Walter the Educator™
> at WaltertheEducator.com

www.ingramcontent.com/pod-product-compliance
Lightning Source LLC
LaVergne TN
LVHW051959060526
838201LV00059B/3738